ANNULAR SOLAR ECLIPSES

GUIDE

Unmasking the Beauty of

Celestial Shadows

SEAN T. RONLAND

COPYRIGHT

Copyright©2024 Sean T. Roland. All rights reserved. No part of this publication may be reproduced, distributed, or transmitted in any form or by any means, including photocopying, recording, or other electronic or mechanical methods, without the prior written permission of the publisher, except in the case of brief quotations embodied in critical reviews and certain other non-commercial uses permitted by copyright law

TABLE OF CONTENTS

ANNULAR SOLAR ECLIPSES GUIDE 0

COPYRIGHT 1

TABLE OF CONTENTS 2

INTRODUCTION 4

    Annular Solar Eclipses 4

CHAPTER 1 6

    Understanding Annular Solar Eclipses 6

CHAPTER 2 12

    The Astronomical Phenomenon: How It Works 12

CHAPTER 3 19

    The History of Annular Solar Eclipses 19

CHAPTER 4 31

    How to Safely Observe an Annular Solar Eclipse 31

CHAPTER 5 42

    The Path of Annularity: Understanding the Shadow 42

CHAPTER 6 52

  Photographing an Annular Solar Eclipse 52

CHAPTER 7 63

  Planning Your Eclipse Viewing Trip 63

CHAPTER 8 76

  Annular Eclipses and Their Effect on Nature 76

CHAPTER 9 87

  Annular Eclipses in Popular Culture 87

CHAPTER 10 99

  Future Annular Solar Eclipses 99

CONCLUSION 112

  Annular Solar Eclipses – An Extraordinary Celestial Phenomenon 112

INTRODUCTION
Annular Solar Eclipses

An annular solar eclipse is a captivating celestial event that occurs when the Moon passes between the Earth and the Sun, but the three bodies do not align perfectly. As a result, instead of completely covering the Sun, the Moon appears smaller, leaving a brilliant ring of sunlight visible around its edges. This phenomenon is often referred to as the "ring of fire" due to its striking appearance.

Unlike total solar eclipses, where the Sun is entirely obscured, annular eclipses provide a unique opportunity to witness the interplay of light and shadow in a spectacular display. The fleeting nature of these eclipses—lasting only a few minutes—adds to their allure, drawing enthusiasts and casual observers alike to experience the breathtaking sight.

Understanding the mechanics behind annular solar eclipses enriches our appreciation of this awe-inspiring event. The alignment of the Earth, Moon, and Sun is a remarkable demonstration of the cosmic dance that shapes our skies. In this guide, we will explore the science, history, and cultural significance of annular solar eclipses, providing you with the knowledge and tips needed to fully enjoy and safely observe these extraordinary occurrences. Whether you are a seasoned astronomer or a curious newcomer, this guide will prepare you for the next annular solar eclipse and deepen your understanding of the wonders of our universe.

CHAPTER 1
Understanding Annular Solar Eclipses

What is an Annular Solar Eclipse?

An annular solar eclipse is a captivating celestial event that occurs when the Moon passes between the Earth and the Sun, yet its apparent size is smaller than that of the Sun. This occurs due to the elliptical nature of the Moon's orbit, which causes its distance from the Earth to vary. When the Moon is closer to Earth, it appears larger; conversely, when it is farther away, it appears smaller. During an annular solar eclipse, the Moon is at or near apogee, the point in its orbit that is farthest from the Earth.

As the Moon obscures the centre of the Sun, it allows sunlight to spill out around the edges, creating a stunning visual effect known as the "ring of fire." This phenomenon is distinct from a total solar eclipse, where the Moon completely covers the Sun, casting a shadow on Earth and plunging the daytime into darkness for a brief period. Instead, during an annular eclipse, the Sun appears as a bright ring encircling the dark silhouette of the Moon, leaving a captivating and unforgettable impression on those who witness it.

To observe an annular solar eclipse, viewers must be located within the path of annularity—the narrow band on Earth where the annular phase can be seen. Outside this path, observers will experience a partial solar eclipse, where the Moon only partially covers the Sun. The duration of an annular eclipse varies, but it typically lasts only a few minutes, with the "ring of fire" visible for a limited time.

Difference Between Annular, Total, and Partial Eclipses

Understanding the differences between annular, total, and partial eclipses is crucial for fully appreciating these celestial events. Each type of solar eclipse presents a unique viewing experience and is characterized by the extent of the Moon's coverage of the Sun.

Total Solar Eclipse

In a total solar eclipse, the Moon completely covers the Sun, blocking its light and casting a shadow on the Earth. This occurs when the Moon is positioned directly between the Earth and the Sun, and its apparent size is larger than that of the Sun. Observers located within the path of totality experience darkness during the day, often witnessing phenomena such as the solar corona—the Sun's outer atmosphere—becoming visible. The temperature drops, and animals may behave as if it is nighttime. The duration of totality can vary, with the longest possible duration being around 7 minutes and 32 seconds, although most eclipses last much shorter than this.

Partial Solar Eclipse

A partial solar eclipse occurs when only a portion of the Sun is obscured by the Moon. This happens when the Moon is not perfectly aligned between the Earth and the Sun or when the observer is outside the path of annularity or totality. During a partial eclipse, observers see only a "bite" taken out of the Sun, with the Moon appearing to overlap it partially. Unlike total or annular eclipses, a partial eclipse does not produce the dramatic visual effects associated with complete coverage, but it still offers a fascinating opportunity to observe the interplay of light and shadow.

Key Differences

The key differences among these three types of solar eclipses lie in their visual characteristics and the observer's experience. A total solar eclipse provides a moment of complete darkness during the day, accompanied by stunning sights such as the solar corona. In contrast, an annular solar eclipse offers a distinctive "ring of fire" effect, while a partial solar eclipse results in a less dramatic view where only part of the Sun is hidden. Each eclipse type occurs under specific geometrical conditions, making them rare and awe-inspiring events that have captivated humanity throughout history.

The Science Behind the "Ring of Fire"

The phenomenon of the "ring of fire" during an annular solar eclipse is a result of the complex interplay between the Earth, Moon, and Sun, along with the physics of light. To understand how this captivating effect occurs, we must delve into the concepts of distance, size, and light behavior.

Geometry of the Earth-Moon-Sun System

The Earth orbits the Sun at an average distance of approximately 93 million miles (150 million kilometers), while the Moon orbits the Earth at an average distance of about 238,855 miles (384,400 kilometers). The elliptical nature of the Moon's orbit means that its distance from the Earth can vary, leading to different apparent sizes of the Moon in the sky. When the Moon is at apogee—its farthest point from Earth—it appears smaller than the Sun, allowing for the annular effect.

During an annular solar eclipse, the alignment of the three celestial bodies is critical. For the eclipse to occur, the Moon must be directly in line with the Sun from the perspective of the Earth while also being at a distance that makes it appear smaller than the Sun. This precise alignment results in the Moon blocking the Sun's central light but allowing sunlight to escape around the edges.

Light and the Atmosphere

Light travels in straight lines, and when the Sun is partially obscured by the Moon during an annular eclipse, the remaining sunlight creates a distinctive glow. The phenomenon is further enhanced by the Earth's atmosphere, which scatters sunlight, giving the "ring of fire" its bright, ethereal appearance. Observers in the path of annularity see a brilliant, continuous ring, while those outside this path witness a partial eclipse with varying degrees of coverage.

Additionally, atmospheric conditions, such as haze or clouds, can impact the visibility and intensity of the "ring of fire." Clear skies provide the best opportunities for viewing and photography, as the contrast between the dark Moon and the bright ring enhances the visual spectacle. In some cases, the refractive effects of the atmosphere can create additional light phenomena, adding to the overall beauty of the event.

In summary, annular solar eclipses are fascinating astronomical events that showcase the intricate relationships between the Earth, Moon, and Sun. They provide a unique opportunity for viewers to witness the breathtaking "ring of fire," a product of precise geometrical alignment and the physics of light. By understanding the differences between annular, total, and partial eclipses, as well as the science behind this captivating phenomenon, we can deepen our appreciation for these celestial displays. As we await the next annular solar eclipse, we are reminded of the wonders of the universe and the beauty that can be observed when we take a moment to look up at the sky.

CHAPTER 2

The Astronomical Phenomenon: How It Works

Orbital Mechanics and the Moon's Path

The dance of celestial bodies in our solar system is governed by the principles of orbital mechanics, which describe the motion of objects in space under the influence of gravitational forces. In the case of solar eclipses, the Earth, Moon, and Sun engage in a complex interplay of gravitational attraction and orbital paths, resulting in the spectacular phenomena we observe from our planet.

The Earth and Moon System

The Earth orbits the Sun at an average distance of about 93 million miles (150 million kilometers) in an elliptical path, taking approximately 365.25 days to complete one revolution. This orbit defines our calendar year and influences the seasonal changes we experience. The Moon, on the other hand, orbits the Earth at an average distance of about 238,855 miles (384,400 kilometers), taking roughly 27.3 days to complete one orbit. This orbital period is known as the lunar month.

As the Moon travels around the Earth, it also rotates on its axis, completing one rotation in the same time it takes to orbit the Earth. This synchronous rotation means that the same side of the Moon always faces our planet, creating what we call the "near side" and "far side" of the Moon. This unique rotational behavior is a result of gravitational interactions over millions of years.

The Moon's Elliptical Orbit

The Moon's orbit around the Earth is not a perfect circle but rather an ellipse. This elliptical shape causes the Moon's distance from the Earth to vary significantly throughout its orbit, with two key points to note: perigee and apogee. **Perigee** is the point at which the Moon is closest to the Earth, while **apogee** is the point at which it is farthest away.

When the Moon is near perigee, it appears larger in the sky, while at apogee, it appears smaller. This variance in apparent size is crucial in determining the type of solar eclipse that can occur. During an annular solar eclipse, the Moon is near apogee, causing it to appear smaller than the Sun and leading to the distinctive "ring of fire" effect.

Why Annular Eclipses Happen Instead of Total Eclipses

Understanding why annular eclipses occur in place of total eclipses requires a grasp of the geometrical alignment of the Earth, Moon, and Sun, as well as the physical characteristics of their orbits.

The Role of Distance

The key factor that differentiates an annular solar eclipse from a total solar eclipse lies in the distance between the Earth and the Moon at the time of the eclipse. When the Moon is at or near apogee during the alignment, it appears smaller in the sky compared to the Sun, resulting in an annular eclipse. Conversely, if the Moon is closer to Earth, at perigee, it will appear larger, leading to a total eclipse.

The Earth-Moon distance is not fixed, and as the Moon's orbit brings it closer or farther from Earth, it directly impacts the type of solar eclipse that occurs. Since the Moon's orbit is elliptical, there are instances when the Moon's position aligns perfectly with the Sun, resulting in an annular eclipse instead of a total eclipse. This means that during certain eclipses, the apparent size of the Moon is insufficient to fully cover the Sun's disk, creating the annular effect.

Orbital Inclination

Another factor to consider is the inclination of the Moon's orbit. The Moon's orbital plane is tilted at an angle of approximately 5 degrees to the ecliptic plane, which is the plane in which the Earth orbits the Sun. This tilt means that the Moon does not pass directly in front of the Sun during every new moon phase. Instead, eclipses occur only when the Moon is positioned at or near one of the two points where its orbit crosses the ecliptic plane, known as nodes.

If the new moon occurs when the Moon is at a node and the Moon is also at apogee, an annular eclipse will occur. If the new moon occurs closer to perigee at a node, a total eclipse will take place. This intricate interplay of distances and alignments dictates the type of solar eclipse that can be observed from Earth.

Conditions Needed for an Annular Eclipse

Several specific conditions must be met for an annular solar eclipse to occur. Understanding these requirements not only enriches our appreciation of this celestial event but also allows us to anticipate and prepare for future occurrences.

1. Proper Alignment

The first and foremost condition for an annular solar eclipse is the precise alignment of the Earth, Moon, and Sun. For an annular eclipse to happen, the three bodies must be nearly in a straight line, with the Moon positioned directly between the Earth and the Sun. However, this alignment must coincide with the Moon being near apogee in its orbit to ensure it appears smaller than the Sun.

2. Moon's Position in Its Orbit

As discussed, the distance of the Moon from the Earth plays a critical role in determining the type of solar eclipse that occurs. The Moon must be near apogee during the new moon phase for an annular eclipse to take place. This positioning limits the potential for a total eclipse, as the smaller apparent size of the Moon cannot entirely obscure the Sun.

3. Location of the Observer

The location of the observer on Earth is also crucial. For an observer to witness an annular eclipse, they must be positioned within the path of annularity—the narrow band on Earth where the annular phase is visible. This path varies with each eclipse and can stretch across different geographical regions. Observers outside this path will experience a partial solar eclipse instead.

4. Clear Atmospheric Conditions

While not a geometric condition, atmospheric conditions play a vital role in the visibility of an annular eclipse. Clear skies are essential for observers to fully appreciate the beauty of the "ring of fire." Haze, clouds, or other atmospheric disturbances can obscure the view and diminish the visual impact of the eclipse. It is always advisable for viewers to monitor weather forecasts leading up to the event to increase their chances of a successful viewing experience.

5. Safety Precautions

Safety is paramount when observing any type of solar eclipse. Viewing the Sun directly can cause serious eye damage, and this is particularly true during an annular eclipse, where the Sun's bright outer edges are still visible. Proper solar viewing glasses or filters are essential to safely observe the event. These tools allow viewers to experience the eclipse without risking their vision, ensuring that the event is not only memorable but safe as well.

In conclusion, the astronomical phenomenon of annular solar eclipses is a fascinating interplay of orbital mechanics, distance, and alignment. Understanding the dynamics of the Earth-Moon-Sun system enhances our appreciation for these celestial events. By exploring the mechanics behind the Moon's path, the conditions leading to annular eclipses, and the unique characteristics that set them apart from total and partial eclipses, we deepen our understanding of the cosmos.

As we anticipate future annular solar eclipses, we are reminded of the intricate dance of celestial bodies and the beauty that can be observed when we take a moment to look up at the sky. The excitement of witnessing an annular eclipse is not just about the visual spectacle; it is also about engaging with the science that underpins these remarkable events and celebrating our connection to the universe at large.

CHAPTER 3
The History of Annular Solar Eclipses

The phenomenon of an annular solar eclipse has long fascinated humanity, evoking awe and wonder across various cultures and historical periods. Annular solar eclipses, often referred to as "ring of fire" eclipses, occur when the Moon passes between the Earth and the Sun but is too far from the Earth to completely cover the Sun's disk, resulting in a bright ring of sunlight encircling the Moon. This unique alignment and its striking visual effect have left an indelible mark on human history, inspiring scientific curiosity, mythology, and cultural practices.

Famous Annular Eclipses in History

Throughout history, several notable annular eclipses have captured the imagination of people around the world. These events not only influenced the development of astronomy but also had significant social, political, and cultural impacts.

The Eclipse of 1836

One of the famous annular eclipses recorded in modern history occurred on May 15, 1836. This event was notable due to the advancements in scientific instrumentation that allowed for detailed observation. Astronomers used the opportunity to measure the precise paths of celestial bodies, furthering their understanding of orbital mechanics. In Britain, this eclipse was a major public spectacle, with people gathering to view the phenomenon. Newspapers at the time were filled with descriptions and illustrations, reflecting both scientific and public interest.

The 1724 Annular Eclipse

The annular eclipse of April 1, 1724, was witnessed across parts of Europe, including France and Spain. It was extensively documented by European scientists who used the opportunity to conduct experiments related to the Sun's corona and its effects on the atmosphere. The phenomenon was seen by the famous astronomer Jacques Cassini, who contributed to the ongoing debates about the distance between celestial bodies.

The 1994 Eclipse in North America

The annular solar eclipse of May 10, 1994, was widely observed across North America, stretching from the Pacific Northwest to the Eastern United States. This event was unique because of the technology available at the time—television broadcasts and live feeds allowed millions to watch the eclipse from the comfort of their homes. The 1994 eclipse also spurred increased interest in astronomy in the United States, with schools and universities using it as an educational tool to teach students about celestial events.

The "Ring of Fire" of 2010

On January 15, 2010, the longest annular solar eclipse of the millennium took place. The event lasted for over 11 minutes, with the path of annularity stretching across Africa, the Indian Ocean, and parts of Asia. It captivated millions, with people from Uganda to India gathering to witness this natural wonder. In India, various religious rituals were conducted in relation to the eclipse, and scientists used this extended event to gather significant data about solar radiation and atmospheric changes.

The Annular Eclipse of 2023

Another well-documented annular eclipse occurred on October 14, 2023, visible across the United States, Central America, and parts of South America. This event became a cultural phenomenon, as people gathered in large numbers with eclipse glasses and pinhole projectors, celebrating the marvel of astronomy. Scientists took this opportunity to study the Sun's corona and to test advanced solar telescopic equipment. The eclipse also highlighted the growing influence of social media in sharing the beauty of natural phenomena, with countless pictures and videos circulating globally.

Ancient Civilisations and Their Eclipse Interpretations

Annular solar eclipses were significant in many ancient civilisations, where they were often perceived as powerful omens, imbued with both religious and social significance. These interpretations shaped cultural attitudes toward eclipses, blending celestial observation with mythology and spiritual beliefs.

The Maya and Eclipses

The Maya civilisation, one of the most advanced pre-Columbian cultures in astronomy, paid close attention to solar and lunar eclipses, including annular eclipses. The Maya viewed eclipses as events that affected the harmony of the cosmos. They believed that celestial alignments were manifestations of divine influence. The Dresden Codex, one of the few surviving Maya texts, contains eclipse tables that demonstrate their capacity to predict such events. The Maya believed that eclipses were times when powerful gods fought, and rituals were performed to protect the Earth from harm during these celestial occurrences.

Ancient China

In ancient China, eclipses were often considered heavenly omens. The Chinese believed that eclipses, including annular ones, signified the displeasure of the heavens, particularly regarding the behaviour of rulers. During the Shang and Zhou Dynasties, records of eclipses were meticulously kept by court astronomers. According to ancient Chinese belief, a dragon would attempt to devour the Sun during an eclipse. To ward off the dragon, people would bang drums and make loud noises to scare it away. There is a famous legend involving two royal astronomers, Hsi and Ho, who failed to predict an eclipse accurately and were subsequently executed, reflecting the importance placed on the interpretation of these events.

Ancient Mesopotamia

The people of Mesopotamia, including the Babylonians, kept some of the earliest known records of eclipses. They meticulously charted celestial phenomena, which played an essential role in their astrological traditions. The Babylonians saw eclipses as omens that could predict the future, particularly concerning the fate of kings. An annular solar eclipse might signify turmoil or change, prompting rulers to perform sacrifices and engage in rituals to mitigate any perceived threats. The priests, known as astrologers, used the information from eclipses to advise kings and determine auspicious times for political or military actions.

Ancient Greece

In Ancient Greece, solar eclipses, including annular eclipses, were seen with a mix of fear and curiosity. Initially, eclipses were interpreted as negative omens, reflecting divine displeasure or signalling catastrophic events. However, as Greek astronomy advanced, these views began to change. The philosopher Anaxagoras was among the first to suggest that eclipses were natural phenomena rather than acts of the gods. The famed astronomer Hipparchus used eclipses to make significant contributions to the understanding of celestial mechanics, including measuring the distance to the Moon. The Greeks transitioned from superstition to science, viewing eclipses as opportunities to study the heavens.

The Egyptians and Solar Eclipses

The ancient Egyptians had a deep reverence for the Sun, personified as the god Ra. Solar eclipses were believed to be times when Ra was under attack by the serpent Apophis. During an eclipse, the struggle between light and darkness became evident, and priests performed rituals to assist Ra in his victory over chaos. However, the Egyptians did not leave extensive records about specific annular eclipses, possibly due to the scarcity of these events or because they perceived them as bad omens and thus sought to avoid them.

Myths, Legends, and Cultural Beliefs

Annular solar eclipses have given rise to a rich tapestry of myths, legends, and cultural practices that illustrate humanity's enduring fascination with the cosmos. These stories offer insight into how different cultures interpreted and responded to the mysterious sight of the Sun being obscured by the Moon, leaving behind a glowing ring.

The "Ring of Fire" Mythology

The striking visual effect of an annular eclipse, often described as a "ring of fire," inspired various myths across cultures. In Hindu mythology, eclipses are linked to the legend of Rahu, a demon who sought immortality by drinking the nectar of the gods. The Sun and the Moon exposed Rahu, and Vishnu beheaded him before he could swallow it. Rahu's disembodied head is said to chase the Sun and the Moon in revenge, causing eclipses when he swallows them momentarily, only for the celestial bodies to reappear since Rahu has no body to hold them.

The Norse Myth of Skoll and Hati

In Norse mythology, solar eclipses are attributed to two wolves, Skoll and Hati. Skoll chases the Sun, while Hati chases the Moon. An annular eclipse was believed to represent a moment when Skoll almost caught the Sun but not entirely, resulting in the "ring of fire." The Vikings would create as much noise as possible—banging shields, yelling, and using horns—to scare away the wolves and save the Sun from being completely devoured.

Native American Beliefs

Different Native American tribes had unique interpretations of eclipses. The Ojibwe, for instance, believed that an eclipse occurred because the Sun and the Moon were momentarily at odds, requiring people to mediate by offering prayers and engaging in rituals to restore harmony. The Choctaw viewed eclipses as a form of reconciliation, a time when the Sun and Moon were "holding hands." In the case of annular eclipses, the "ring of fire" was seen as a symbol of the cosmic connection and unity that should be reflected in human relationships.

African Eclipse Myths

In various African cultures, solar eclipses, including annular ones, were seen as omens that signified significant changes or warnings. Among the Batammaliba people in Togo and Benin, an eclipse is viewed as a conflict between the Sun and the Moon. The people interpret this cosmic quarrel as a sign that they too must resolve their conflicts, using the occasion to foster peace within their communities. The annular eclipse, with its distinct visual pattern, was particularly associated with the idea of a "celestial ring," symbolising unity and wholeness, encouraging collective harmony.

Japanese Eclipse Folklore

In Japanese mythology, eclipses were often linked to notions of impurity and imbalance. The annular eclipse was sometimes seen as a celestial warning, prompting purification rituals. The Japanese would cover wells and hide water sources to prevent contamination during an eclipse, reflecting the belief that these events temporarily disrupted the natural order. In Japanese Shinto practices, annular eclipses were moments of reflection on cosmic balance, and priests would perform ceremonies to appease the kami (spirits) and restore harmony.

Eclipses in Islamic Tradition

In Islamic tradition, eclipses are regarded as signs of Allah's greatness, devoid of superstitions or myths commonly found in other cultures. The Prophet Muhammad emphasised that eclipses were natural events and instructed people to offer special prayers, called Salat al-Khusuf, during these occurrences. The annular eclipse, with its breathtaking "ring of fire," is seen as a moment to reflect on the majesty of creation. Rather than fear, it inspires a sense of humility and reverence for the divine.

Modern Cultural Beliefs

In contemporary times, annular solar eclipses have become opportunities for scientific education and cultural gatherings. Many people around the world travel great distances to witness an annular eclipse, treating it as a shared human experience that transcends borders. Festivals, viewing parties, and educational workshops are often organised around the path of annularity, with people using specialised glasses and projection devices to safely view the phenomenon. This modern approach has transformed what was once a source of fear and superstition into an occasion for celebration and learning.

The Psychological Impact of Eclipses

The visual spectacle of an annular eclipse has a psychological impact that can evoke a deep sense of connection to the universe. Many people describe feelings of awe and even spirituality when witnessing the "ring of fire." The temporary obscuration of the Sun disrupts the normal perception of day and night, creating an otherworldly atmosphere that can lead individuals to reflect on their place in the cosmos. This effect has been documented across cultures, where the communal experience of an eclipse often leads to shared rituals, fostering a sense of unity.

Scientific Shifts and Cultural Continuity

While the scientific understanding of annular eclipses has grown substantially, many cultural practices and beliefs have persisted. The juxtaposition of science and myth continues to characterise humanity's relationship with celestial phenomena. In some cultures, traditional rituals are still performed, even as scientific explanations are accepted. The duality of interpreting an eclipse both as a "ring of fire" shaped by cosmic mechanics and as a spiritual symbol highlights the richness of human culture, where science and mythology coexist, enriching the experience of observing such awe-inspiring events.

Annular solar eclipses have thus woven themselves into the history and culture of humanity in myriad ways. From the fearful cries of ancient civilisations to the cheers of modern eclipse chasers, the "ring of fire" has always ignited a sense of wonder and reflection, reminding us of both our smallness in the vast universe and our capacity for curiosity and understanding. The fusion of myth, culture, and science in the history of annular eclipses reveals much about how humanity perceives its place in the cosmos—an intricate dance of reverence, curiosity, and the pursuit of knowledge.

CHAPTER 4

How to Safely Observe an Annular Solar Eclipse

The breathtaking beauty of an annular solar eclipse, often called the "ring of fire," can be an unforgettable experience, but it's essential to prioritise safety when observing it. Unlike a total solar eclipse, where the Sun is entirely blocked, an annular eclipse leaves a bright ring of sunlight visible, meaning there is never a point during the event where it is safe to look at the Sun without proper protection. The light is still intense enough to cause serious damage to your eyes if viewed without proper tools or precautions. Below, we'll explore the different methods and tools to observe this natural wonder safely, what equipment you should use, and common mistakes to avoid.

Eye Safety: Viewing Methods and Tools

Observing an annular solar eclipse safely requires an understanding of the risks associated with looking directly at the Sun and the methods to mitigate these dangers. The Sun's rays contain intense ultraviolet (UV) and infrared (IR) radiation that can cause severe damage to the retina, potentially leading to permanent vision loss, a condition known as solar retinopathy. The following are recommended methods to ensure safe viewing:

Eclipse Glasses

The most straightforward and popular method for safely viewing an annular eclipse is using certified solar eclipse glasses. These glasses are designed with special-purpose solar filters that block out harmful rays, allowing you to view the Sun without risking your eyesight. It's crucial to ensure that your eclipse glasses comply with the ISO 12312-2 international safety standard, which guarantees that the lenses have adequate protection against ultraviolet, infrared, and intense visible light. When purchasing eclipse glasses, always check for the certification label, as counterfeit products may not provide the necessary level of protection.

Pinhole Projectors

A pinhole projector is an easy and effective way to observe an eclipse indirectly. This method involves using a piece of cardboard or a similar material with a small hole in it. You allow the sunlight to pass through the hole, projecting an image of the Sun onto a surface, such as a sheet of white paper. This approach is entirely safe because you are not looking directly at the Sun. Pinhole projectors are particularly popular for group settings and with children, as they are easy to make and allow people to watch the eclipse without any risk.

Projection Using Binoculars or a Telescope

If you have a pair of binoculars or a small telescope, you can use them to project an image of the eclipse onto a screen or a piece of paper. To do this, point the binoculars or telescope at the Sun while standing behind it and adjust until you see a clear image of the Sun projected onto your surface. It's important to note that you should never look through the binoculars or telescope directly unless they are fitted with solar filters specifically made for safe solar observation. This projection method is advantageous as it provides a larger image of the eclipse, making it easier for a group of people to view.

Welders' Glass

Welders' glass rated at a shade level of 14 or higher can also be used to observe an annular eclipse. This type of glass blocks out harmful radiation and reduces the intensity of sunlight to a safe level for direct viewing. However, it's essential to confirm the glass's rating, as lower-level welding glasses do not provide sufficient protection. This method is often recommended for those who want a quick and readily available alternative to eclipse glasses, though it may not be as comfortable for prolonged viewing.

Special Solar Viewers and Filters

Special solar viewers and filters, often used by amateur astronomers, offer another safe method of eclipse observation. These viewers are mounted on telescopes or binoculars, allowing for magnified views of the annular eclipse without the associated dangers. Make sure these solar filters are fitted to the front end of the equipment, not at the eyepiece, as the concentrated solar rays can damage both the filter and the equipment if improperly positioned. High-quality solar filters are made from materials such as black polymer, aluminised polyester film, or optical glass designed for safe solar viewing.

Choosing the Right Equipment: Solar Filters and Glasses

Selecting the proper equipment for viewing an annular solar eclipse is vital to protect your eyesight. Here are some options for solar filters and glasses, along with tips on how to ensure you have the safest tools:

Solar Filters for Telescopes, Binoculars, and Cameras
When viewing an annular solar eclipse through a telescope or binoculars, it is imperative to use solar filters explicitly

designed for solar observation. These filters should be installed at the front end of the optical device, as they block out a significant portion of the Sun's radiation before it even enters the telescope or binoculars. Using an eyepiece filter can lead to concentrated solar radiation damaging the equipment and potentially causing injury.

When choosing a solar filter, look for products from reputable manufacturers that comply with international safety standards. The solar filter should be securely attached to ensure it does not fall off during observation. Some filters are designed to fit specific models of telescopes and binoculars, so it is important to select a filter that matches your equipment.

Solar Eclipse Glasses

Not all eclipse glasses are created equal. With the surge of interest during eclipse seasons, counterfeit glasses with inadequate protection have become more common. To choose the right pair of solar eclipse glasses:

1. **Certification**: Look for the ISO 12312-2 label, which certifies that the glasses meet international safety standards for direct solar observation.

2. **Check for Damage**: Before using your glasses, inspect them for any scratches, tears, or other damage. Even small defects can significantly reduce the level of protection and may allow harmful solar radiation through.

3. **Reputable Sellers**: Purchase glasses from reputable manufacturers or astronomy organisations. Brands such as American Paper Optics, Rainbow Symphony, Thousand Oaks Optical, and Lunt Solar Systems are known for producing high-quality solar viewing products.

4. **No Use with Regular Sunglasses**: Regular sunglasses, even those with high UV protection, are not safe for viewing an eclipse. They do not block enough of the Sun's light and radiation to be used during an eclipse.

Welders' Glass

Welders' glass rated at shade 14 is a practical option, especially if eclipse glasses are not available. The shade level is crucial: anything less than 14 does not provide adequate protection. When using welders' glass, check that it is free from scratches or defects, as these can compromise its effectiveness. Welders' glass is often heavier and darker than eclipse glasses, which can make it less comfortable for extended viewing sessions, but it remains a reliable and accessible option.

Solar Filters for Cameras and Smartphones

With the widespread use of smartphones for documenting celestial events, it's essential to note that pointing your camera directly at the Sun without proper solar filters can damage the camera sensor. Solar filters designed for camera lenses are available, allowing safe photography during an eclipse. These filters must be securely attached to prevent accidental exposure. Using a smartphone to take pictures of an eclipse should be done indirectly, such as by photographing the projected image from a pinhole projector or using solar-filtered equipment.

What Not to Do During an Eclipse

While the excitement surrounding an annular eclipse can be overwhelming, it's critical to avoid certain behaviours that could lead to serious injury or equipment damage. Below are the key things to avoid:

Never Look Directly at the Sun Without Protection
This is perhaps the most critical rule when observing an annular eclipse. The Sun's rays are extremely powerful, and

looking directly at the eclipse without proper eye protection—even for a brief moment—can cause irreversible damage to your retina. This damage, known as solar retinopathy, is painless, so you might not realise the extent of the injury until later. Always use certified eclipse glasses or solar viewing tools when looking directly at the Sun.

Avoid Using Regular Sunglasses

Regular sunglasses, even those with 100% UV protection, are not safe for viewing an eclipse. They do not filter out enough of the Sun's harmful rays to prevent eye damage. Even multiple pairs of sunglasses stacked together are inadequate. Only solar viewing glasses certified for eclipse viewing should be used.

Do Not Use Tinted Glass, CDs, or Other Improvised Filters

Some people attempt to use tinted glass, CDs, DVDs, or other household items as makeshift filters to view an eclipse. These items are dangerous and do not provide adequate protection from the Sun's intense radiation. They may reduce the brightness but do not block the harmful UV and IR rays that can damage your eyes. Always opt for professionally manufactured solar filters and eclipse glasses.

Never Look Through Binoculars, Telescopes, or Cameras Without Solar Filters

Using binoculars, telescopes, or even cameras to view an eclipse without appropriate solar filters is extremely dangerous. These devices amplify sunlight, focusing it directly into your eyes, which can cause severe burns to the retina almost instantly. The use of any optical device to view an eclipse requires that solar filters are installed at the front end to block the light before it is magnified. This precaution applies not only to telescopes and binoculars but also to any photographic equipment being used to capture the event.

Avoid Removing Solar Glasses While Looking at the Sun
One common mistake people make is removing their solar glasses while the eclipse is still in progress. The "ring of fire" during an annular eclipse is still too bright to look at without protection. You should only remove your solar glasses when you are not facing the Sun. It's easy to underestimate the brightness during this phase, but even a small portion of visible sunlight is enough to damage your eyes.

Do Not Rely on Improperly Secured Filters
Whether you are using a solar filter for a telescope or binoculars, ensure that it is securely fastened. A loose filter can fall off unexpectedly, exposing your eyes to direct

sunlight through magnifying lenses. This exposure can cause immediate injury. Double-check all equipment before the eclipse starts to confirm that everything is securely in place.

Do Not Underestimate the Dangers of Infrared and Ultraviolet Radiation

While the visible light from an eclipse is dazzling, the unseen ultraviolet and infrared radiation can be even more damaging. These forms of radiation are invisible, so the danger is often overlooked. Proper solar filters are designed to block both UV and IR radiation, ensuring complete protection during viewing. It's critical not to use equipment or viewing methods that only reduce visible light but do not address UV and IR exposure.

Avoid Leaving Children Unsupervised During the Eclipse

Children are often excited by the prospect of an eclipse but may not understand the dangers of looking directly at the Sun. It's crucial to supervise children during the event, ensuring they keep their eclipse glasses on while viewing. Providing indirect viewing methods, like pinhole projectors, is a safer option for young children, allowing them to participate without risking accidental exposure.

Don't Ignore Weather Conditions

Cloud cover can sometimes obscure an eclipse, but even when the Sun appears dimmed behind clouds, it is still dangerous to look directly at it without proper protection. The filtered sunlight can create an illusion that it is safe to view, leading to potential eye damage. Always use solar glasses or other safe methods, regardless of weather conditions.

Observing an annular solar eclipse can be a magnificent and awe-inspiring experience, but it requires careful preparation to ensure safety. By following these guidelines, using the right equipment, and avoiding common mistakes, you can safely enjoy the wonder of the "ring of fire." Remember, eye safety should always be the top priority, as the consequences of improper viewing can be permanent and severe. Taking appropriate precautions not only protects your eyesight but also allows you to fully appreciate the beauty of this remarkable celestial event.

CHAPTER 5

The Path of Annularity: Understanding the Shadow

Annular solar eclipses, known for their spectacular "ring of fire" effect, occur when the Moon passes in front of the Sun, but due to its distance from the Earth, it appears slightly smaller and leaves the outer rim of the Sun visible. This results in a glowing halo—a visual spectacle cherished by astronomers and casual observers alike. One of the critical aspects of experiencing an annular eclipse is understanding the path of annularity, which defines where the event is visible in its full form. Below, we explore how to predict the path of annular eclipses, the factors that affect visibility, and the tools available to help enthusiasts plan for upcoming events.

Predicting the Path: Where to See Annular Eclipses

The "path of annularity" refers to the specific region of the Earth where an annular solar eclipse can be seen in its full glory—where the Moon passes centrally across the Sun, creating the ring of fire effect. Predicting this path requires a precise understanding of the movements of both the Earth and the Moon, taking into account their orbital mechanics and alignment with the Sun.

Celestial Mechanics and Predicting the Path

The path of an annular solar eclipse is calculated based on the orbits of the Earth and the Moon around the Sun. The Moon's orbit around the Earth is elliptical, which means that at times it is closer to the Earth (perigee) and at other times farther away (apogee). An annular eclipse occurs when the Moon is at or near its apogee, making it appear slightly smaller in the sky than the Sun, resulting in the ring of fire effect during maximum eclipse.

Astronomers use complex mathematical models to predict the path of annularity years, sometimes even decades, in advance. These calculations involve the relative positions of the Sun, Moon, and Earth, as well as their gravitational interactions. The path of annularity is typically a few hundred kilometres wide and can extend for thousands of kilometres across the Earth's surface. This narrow path is where viewers can see the annular phase, while those outside it, but within the broader shadow, see only a partial eclipse.

Regions Along the Path

The path of annularity typically begins at sunrise on one part of the Earth, travels across continents or oceans, and ends at sunset on the opposite side of the globe. The path varies significantly depending on the geographical location and the time of year. The regions within this path are the only places where observers can witness the complete annular phase, while others located outside the path may experience only a partial eclipse.

For example, the annular eclipse of October 14, 2023, had a path that began in the Pacific Ocean, crossed North America through states such as Oregon, Nevada, and Texas, and then continued through Central America before ending in the Atlantic Ocean. By understanding the predicted paths, observers can plan their travels to locations where they can witness the eclipse in its full annular form.

Planning to See an Annular Eclipse

Predicting the path of an annular eclipse allows individuals and communities to prepare in advance to witness this celestial event. Astronomical organisations such as NASA, the European Space Agency (ESA), and others provide detailed predictions of eclipse paths, including maps and exact timings, which allow observers to determine the best locations to see the event. Travelling to the path of annularity is often required for the best view, and "eclipse chasing" has become a popular activity among enthusiasts who plan vacations and expeditions around these rare events.

Factors Affecting Visibility: Weather, Altitude, and Location

While understanding the path of annularity is critical, several other factors play a role in determining whether one will have a successful viewing experience. Weather conditions, altitude, and specific geographic location can all significantly impact visibility during an eclipse.

Weather Conditions

Weather is one of the most unpredictable factors that can affect eclipse visibility. Cloud cover can obstruct the view of the Sun entirely, leading to a disappointing experience for observers. Therefore, it is essential to monitor weather forecasts in the days leading up to the eclipse. Locations with a historical record of clear skies are often chosen for eclipse viewing.

For instance, areas with desert climates, such as parts of the American Southwest, are popular for eclipse viewing because they tend to have fewer clouds compared to tropical regions. For those planning to travel to witness an annular eclipse, it's advisable to have a backup location within reasonable travel distance to increase the chances of clear weather.

Altitude

The altitude of the viewing location can also affect the visibility of an annular eclipse. Higher altitudes tend to offer clearer views because there is less atmospheric interference, such as dust or haze. Mountainous regions are often ideal for observing eclipses, provided they are within the path of annularity, as the thinner atmosphere reduces the distortion of sunlight.

High-altitude viewing, however, comes with its own set of challenges, such as accessibility and the potential for extreme weather conditions. Proper planning and preparation are necessary to ensure that viewers are safe and comfortable while experiencing the eclipse.

Geographic Location

Geographic location plays a crucial role in determining whether an observer will see the eclipse as annular or partial. The duration of the annular phase also varies depending on where along the path of annularity an observer is located.

The closer one is to the central line of the path, the longer the duration of the annular phase.

Locations near the edge of the path may only experience a brief annular phase or even miss it altogether, resulting in a partial eclipse instead. Viewing from the centre of the path ensures the longest duration of the ring of fire effect, which can last anywhere from a few seconds to over 10 minutes, depending on the specific eclipse.

Local Terrain and Horizon

The local terrain can also impact visibility. Viewing from an open area with a clear horizon is ideal, especially if the eclipse occurs close to sunrise or sunset. Mountains, buildings, and trees can obstruct the view, making it harder to see the full progression of the eclipse. Choosing a site with an unobstructed view of the horizon is particularly important for capturing the beginning or end of the eclipse.

Maps and Charts for Upcoming Annular Eclipses

Detailed maps and charts are indispensable tools for planning to observe an annular solar eclipse. These resources provide information about the path of annularity, including the geographic areas that will experience the full annular phase, the duration of annularity, and the best times to observe.

Eclipse Maps

Maps of the path of annularity are typically produced by astronomical organisations and are freely available online. These maps show the route that the eclipse will take across the Earth's surface, indicating the regions that will fall within the path of annularity. Maps often include additional details such as the time of maximum eclipse, the duration of annularity, and the locations that will experience a partial eclipse.

Websites such as NASA's Eclipse Page, TimeandDate.com, and the International Astronomical Union (IAU) provide interactive maps that allow users to zoom in on specific regions and determine the exact timing and duration of the eclipse at their location. These tools are invaluable for anyone planning to travel to view an eclipse, as they help identify the most promising viewing spots along the path.

Upcoming Annular Eclipses

One of the most anticipated upcoming annular eclipses will occur on October 2, 2024. The path of annularity for this eclipse will pass through parts of South America, including

Chile and Argentina, before moving across the Atlantic Ocean. This event is expected to draw eclipse chasers from around the world, as the South American Andes provide an opportunity for high-altitude viewing, enhancing the overall experience.

Another notable annular eclipse will take place on February 17, 2026. The path of annularity will cross parts of Africa, including regions of Angola and Zambia, before moving through the Indian Ocean. This eclipse will offer a unique opportunity to view the ring of fire in remote and less frequented locations, allowing for an intimate viewing experience with minimal crowd interference.

Using Eclipse Charts

Eclipse charts are another useful resource for planning. These charts provide detailed information about the timings of the eclipse, including when the partial phases begin, when the annular phase starts and ends, and the duration of each phase. The timing of an eclipse is typically given in Coordinated Universal Time (UTC), so observers need to convert these times to their local time zones.

Charts often include information about the altitude and azimuth of the Sun during each phase of the eclipse, which is helpful for determining where in the sky the eclipse will occur. This allows viewers to plan their positioning and equipment setup in advance, ensuring that they have a clear line of sight to the Sun throughout the event.

Mobile Apps and Digital Tools

In recent years, mobile apps have become popular tools for planning and observing eclipses. Apps like Solar Eclipse Timer and Eclipse Guide provide users with detailed information about upcoming eclipses, including the path of annularity, weather forecasts, and viewing tips. These apps can also send notifications during the eclipse to guide users through the different phases, helping them make the most of their viewing experience.

Digital tools and interactive maps offer an immersive way to explore the path of upcoming annular eclipses. Many websites allow users to enter their location and receive specific details about what to expect during the eclipse, including the percentage of the Sun that will be covered and the exact timing of each phase. These resources are particularly useful for individuals who may not be able to travel to the path of annularity but still want to experience a partial eclipse.

Global Eclipse Viewing Events

Public viewing events are often organised in locations along the path of annularity. These events provide a safe and educational environment for observing the eclipse, with experienced astronomers on hand to explain the phenomenon and answer questions. Many events provide access to telescopes equipped with solar filters, allowing participants to get a closer view of the eclipse. Attending a public viewing event can enhance the experience, providing opportunities for learning and for sharing the excitement of an eclipse with others.

Annular solar eclipses offer a rare and extraordinary opportunity to witness the dynamic interactions between the Earth, Moon, and Sun. Understanding the path of annularity is key to planning a successful viewing experience, but it's also essential to consider other factors such as weather, altitude, and local terrain. By utilising maps, charts, and digital tools, eclipse enthusiasts can prepare thoroughly and position themselves in the best possible locations to experience the spectacle of the "ring of fire." Whether you are a seasoned eclipse chaser or a first-time viewer, careful planning and preparation can ensure a safe and memorable encounter with this celestial wonder.

CHAPTER 6

Photographing an Annular Solar Eclipse

Photographing an annular solar eclipse is an exciting challenge for both amateur and professional photographers. Capturing the "ring of fire" is a rewarding experience, but it requires careful preparation and the right techniques. Given the unique conditions of the event, understanding the proper photography gear, techniques for capturing the annular phase, and common pitfalls to avoid is key to producing stunning eclipse photographs.

Essential Photography Gear for Eclipse Photography

To successfully capture an annular solar eclipse, having the right photography equipment is essential. The sunlight during the annular phase is still powerful enough to damage cameras and eyes, so using appropriate gear is critical for both safety and getting high-quality results.

1. Camera

While it is possible to photograph an eclipse with various types of cameras, a DSLR or mirrorless camera is ideal due to its flexibility, manual controls, and the ability to change lenses. These cameras allow you to control exposure, shutter speed, ISO, and focus more precisely than a point-and-shoot camera or a smartphone.

However, modern smartphones equipped with high-quality lenses can also be used to capture the event, particularly when using external lenses or with the assistance of a telescope. For best results, a camera capable of shooting in RAW format is preferable, as it provides more flexibility for post-processing.

2. Telephoto Lens

A telephoto lens is crucial for capturing close-up images of the annular solar eclipse. A lens with a focal length of at least 400mm is recommended, as this provides a tight frame of the Sun, allowing you to capture the intricate details of the ring of fire. A longer focal length, such as 600mm or even 800mm, will provide more detail and help fill the frame with the Sun.

If you don't have access to a long telephoto lens, you can also use a telescope as a camera lens with the help of an adapter, which can provide an effective focal length of several thousand millimetres.

3. Solar Filter

A solar filter is arguably the most critical piece of gear for photographing an eclipse. The solar filter should be fitted to the front of the lens to protect both the camera sensor and your eyes from the Sun's intense rays. Solar filters reduce the brightness of the Sun and eliminate harmful infrared and ultraviolet radiation, allowing you to safely photograph the eclipse.

Ensure that the solar filter is securely attached to the camera or telescope, as an improperly fitted filter could fall off, resulting in permanent damage to the camera and potential injury. There are different types of solar filters available, such as glass or mylar, each offering slightly different characteristics. Make sure to use a high-quality solar filter that meets safety standards.

4. Tripod

Stability is key when photographing a solar eclipse. A sturdy tripod is essential for keeping the camera steady, especially since you will be using a long focal length, which can amplify even the slightest vibrations. A ball-head or fluid-head tripod will allow for more flexibility in positioning the camera, enabling you to track the Sun smoothly as it moves across the sky.

5. Remote Shutter Release

A remote shutter release allows you to take photos without physically touching the camera, which helps reduce vibrations that could cause blurring. It is especially useful for eclipse photography, as you may want to capture multiple shots over an extended period without introducing any movement to the camera. Alternatively, the camera's built-in timer can be used, but a remote release offers more convenience and control.

6. ND Filter (Optional)

In addition to a solar filter, a neutral density (ND) filter can help reduce the overall brightness of the scene, especially during the partial phases of the eclipse. ND filters are less effective than dedicated solar filters for direct Sun photography but can be used creatively to capture the ambient environment during the eclipse.

7. Extra Batteries and Memory Cards

A solar eclipse can last for several hours, and you will likely be taking numerous photos throughout the event. Having extra batteries and memory cards on hand is important to ensure you do not run out of power or storage during critical moments.

Techniques for Capturing the "Ring of Fire"

Successfully photographing an annular eclipse, and especially capturing the iconic ring of fire, requires proper planning and a good grasp of exposure settings. Below are some techniques to help achieve stunning images:

1. Use Manual Settings

The key to photographing an eclipse is using manual settings. Automatic modes can struggle to deal with the brightness of the Sun against the dark sky, leading to incorrect exposure. Start with the following settings:

- **Shutter Speed**: A shutter speed of 1/500 to 1/1000 of a second is often a good starting point, as the Sun is extremely bright.

- **Aperture**: Use a medium to small aperture, such as f/8 to f/16, to get a good depth of field and minimise chromatic aberrations.

- **ISO**: Keep the ISO as low as possible (typically ISO 100) to reduce noise in your images.

These settings may need adjustment depending on the brightness of the Sun and your specific location, but they offer a good starting point for capturing clear images.

2. Focus Manually

Autofocus often struggles during an eclipse because the camera may have difficulty finding the correct focus point. Switch to manual focus and use the live view function to zoom in on the edge of the Sun and achieve precise focus. Make sure to focus before the main phase of the eclipse begins, as you may not have time to adjust once the event is underway.

3. Bracketing Exposures

To ensure that you capture a range of details in your photographs, consider bracketing your exposures. This involves taking multiple shots at different exposures to capture varying levels of light and detail. Since the lighting conditions change throughout the eclipse, bracketing ensures that you will have at least some properly exposed images to work with later.

4. Time-lapse Photography

Time-lapse photography is an excellent way to capture the entire progression of the eclipse, from the partial phases to the ring of fire and back again. Set your camera to take photos at regular intervals, such as every 30 seconds or one minute. This approach will allow you to compile the images into a time-lapse video that shows the entire eclipse in a short sequence, providing a dramatic visual representation of the event.

5. Composition and Environment

While it's easy to focus solely on capturing the Sun, incorporating elements of the surrounding landscape can create more compelling images. If possible, position yourself in a location where you can include foreground features like mountains, trees, or structures to give a sense of scale and context. This technique works particularly well when the Sun is lower on the horizon during the start or end of the eclipse.

6. Test Shots Before the Eclipse

Practicing and preparing ahead of time are key to capturing great images. Take test shots of the Sun using your solar filter a few days before the eclipse. This will help you get comfortable with your camera settings, ensure that your solar filter is functioning correctly, and allow you to troubleshoot any problems before the big day.

Common Mistakes and How to Avoid Them

Photographing an annular eclipse is an exciting opportunity, but there are several common mistakes that can easily be made if not properly prepared. Here are the most frequent pitfalls and tips for avoiding them:

1. Not Using a Solar Filter

Failing to use a solar filter is perhaps the most dangerous mistake a photographer can make during an eclipse. Directly pointing your camera at the Sun without proper filtration can permanently damage the camera sensor and potentially injure your eyes. Always use a certified solar filter to protect both your equipment and yourself.

2. Overexposing or Underexposing

The Sun is extremely bright, even during an eclipse, and incorrect exposure can lead to washed-out images or photos that are too dark to reveal detail. Start with manual settings and adjust based on your observations, and consider bracketing your exposures to capture a variety of brightness levels. Using the histogram on your camera can also help you evaluate whether your shots are properly exposed.

3. Camera Shake and Blurred Images

Blurry images can result from camera shake, which is particularly problematic when using long focal lengths. Use a sturdy tripod and a remote shutter release to avoid any unintentional vibrations. If you don't have a remote, use the camera's timer function to eliminate shake caused by pressing the shutter button.

4. Forgetting to Track the Sun's Movement

The Sun moves quite quickly across the sky, so you need to adjust your camera's position frequently to keep it in the frame. Many photographers forget this and end up with images where the Sun has drifted out of view. Make sure to regularly reposition your camera and tripod to keep the Sun centred.

5. Relying on Autofocus

Autofocus can be unreliable when photographing an eclipse due to the brightness of the Sun and the lack of contrast in the image. It's best to switch to manual focus and use live view to zoom in and set the focus manually before the eclipse begins. Practice achieving a sharp focus in advance to avoid struggling during the event.

6. Poor Planning for Location and Timing

Location is critical for successful eclipse photography. Choosing a location without a clear view of the horizon, or failing to consider the weather forecast, can result in a less-than-ideal experience. Scout your location ahead of time, and check weather predictions leading up to the eclipse day.

It's also advisable to arrive well before the eclipse starts so you can set up and make any last-minute adjustments.

7. Forgetting Extra Gear

Solar eclipses are lengthy events, and running out of power or memory in the middle of the action is a common mistake. Bring extra batteries and memory cards to ensure that you are well-prepared to capture the entire sequence. It's better to have more capacity than needed, as you will likely take more shots than initially planned.

8. Ignoring the Surrounding Environment

The eclipse itself is captivating, but often photographers overlook the opportunity to capture the effects of the eclipse on the surrounding environment. The lighting changes dramatically during an eclipse, creating unique shadows and dimming the landscape in an unusual way. Capturing the reaction of onlookers, wildlife behaviour, or the changing colours of the sky can add an extra dimension to your eclipse photography.

9. Not Practicing Before the Main Event

An eclipse is not the time to be learning how to use your gear. Practice beforehand to familiarise yourself with the settings, equipment, and the solar filter. Test out different shutter speeds, apertures, and ISO settings to understand how your camera will respond to the Sun. The more familiar you are with your gear, the more likely you are to capture successful images during the actual eclipse.

Photographing an annular solar eclipse is a memorable and rewarding endeavour, but it requires careful planning, proper equipment, and attention to detail. By understanding the essential gear needed, mastering the right techniques, and avoiding common mistakes, you can create striking images that showcase the stunning beauty of the ring of fire. Whether you are a seasoned photographer or a curious beginner, the experience of capturing an eclipse will be one you won't forget.

CHAPTER 7
Planning Your Eclipse Viewing Trip

An annular solar eclipse is a captivating natural phenomenon, often referred to as the "ring of fire" for the dazzling effect created when the Moon, at its farthest point from Earth, covers most of the Sun but leaves a glowing ring around the edges. For many enthusiasts, viewing an annular eclipse in its full glory requires travelling to specific locations along the path of annularity. This means careful planning, choosing the right location, and considering travel logistics to ensure a successful viewing experience. Below, we cover the best locations worldwide for viewing annular eclipses, travel tips, and an exploration of eclipse tourism.

Best Locations Worldwide for Viewing Annular Eclipses

The path of annularity for any given eclipse is limited to a narrow strip of the Earth's surface, often spanning thousands of kilometres but only a few hundred kilometres wide. Where you can see an annular eclipse depends on the path predicted for that particular event. Each eclipse presents different opportunities, as the path changes based on the orbits of the Earth and Moon. Below are some of the most popular and consistently favourable regions worldwide for viewing annular eclipses.

1. Western United States

The American Southwest is one of the best places for eclipse viewing due to its clear skies and arid climate. States like Nevada, Utah, Arizona, and Texas are popular destinations for observing annular eclipses, as they frequently fall within the path of annularity. The desert climate in these regions usually means minimal cloud cover, providing an unobstructed view of the sky. Additionally, areas like Monument Valley and the Grand Canyon offer stunning landscapes that can add depth to eclipse photography and provide an unforgettable backdrop.

2. Chile and Argentina

South America has hosted several prominent solar eclipses in recent years, and both Chile and Argentina are well-suited for eclipse viewing. The Andes Mountains provide high-altitude locations that minimise atmospheric interference, improving visibility. The dry climate in the Atacama Desert in northern Chile makes it an ideal place to observe an eclipse. Argentina's wide-open plains and fewer obstructions also create perfect viewing conditions.

3. India

India is a frequent destination for eclipse enthusiasts due to its positioning along several past paths of annularity. The country offers a wide range of landscapes, from coastal areas to high-altitude plateaus, giving viewers plenty of choices for eclipse viewing locations. The Rann of Kutch, a vast salt desert in Gujarat, is one such area where a clear horizon and unobstructed view make for perfect eclipse conditions.

4. Oman and Saudi Arabia

For those looking to experience an eclipse in the Middle East, Oman and Saudi Arabia are excellent choices. These countries offer open desert landscapes with minimal light pollution, providing an excellent setting for observing the eclipse. The low probability of cloud cover during the year makes these regions particularly reliable for eclipse viewing. Oman, with its dramatic coastline and mountainous regions, adds a scenic element that makes the experience even more memorable.

5. African Savannas

African countries, such as Tanzania and Kenya, have also hosted annular eclipses in the past. Viewing an eclipse over the savanna, with iconic landscapes and wildlife, adds an element of adventure to the experience. The combination of unique wildlife behaviour during the eclipse and the stunning African sunset can make for a once-in-a-lifetime experience.

6. Indonesia

Indonesia, with its islands stretching across the equator, frequently falls along the path of annular and total eclipses. The country's diverse geography, including beaches, mountains, and jungles, allows for unique viewing opportunities. Indonesian islands such as Sumatra or Java have hosted many eclipse viewing events, where visitors can witness the eclipse in a tropical setting while enjoying the natural beauty of the region.

Travel Tips: Logistics, Accommodation, and Timing

Planning an eclipse viewing trip requires a considerable amount of forethought, particularly since the path of annularity is often located in remote or less accessible regions. From organising logistics and securing accommodation to understanding the timing, here are essential travel tips to make your eclipse trip a success:

1. Choose the Location Early

The first and most important step is selecting the location where you want to view the eclipse. Since the path of annularity can span thousands of kilometres, it's wise to choose a location that maximises the chance of clear skies and has good accessibility. Make this decision well in advance, as popular eclipse viewing spots tend to fill up quickly with fellow enthusiasts and tourists.

Consider a location with minimal cloud cover during the time of year the eclipse is occurring. Historical weather patterns can give you a good idea of what to expect, and websites like TimeandDate.com often provide cloud cover predictions to help you plan effectively.

2. Book Accommodation Well in Advance

Once you've chosen your viewing location, secure accommodation as early as possible. During an eclipse, particularly in popular viewing areas, local hotels, campsites, and lodges can fill up months in advance. If the eclipse is passing through a remote or less developed area, accommodation may be limited, and planning becomes even more crucial.

Camping is often a popular choice for those seeking to view the eclipse in more remote locations, as it allows you to set up directly along the path of annularity and stay away from the crowds. Many eclipse enthusiasts opt to rent vehicles and camp to increase flexibility in case weather conditions change.

3. Timing and Arrival

Aim to arrive at your destination at least one or two days before the eclipse. This allows time to scout for a perfect spot, set up equipment, and make adjustments based on weather conditions. Arriving early also gives you time to familiarise yourself with the area, find backup locations in case of clouds, and test your gear to avoid any technical difficulties.

Consider the time of the eclipse as well. Annular eclipses can occur at different times of the day—early morning, midday, or late afternoon. Be aware of sunrise or sunset times and how they might affect your ability to view or photograph the eclipse. Checking eclipse charts and maps that provide the timing for each phase of the event at your location is crucial for ensuring you are in place and ready well before the start.

4. Local Transportation

The path of annularity often crosses regions with limited transportation options. Renting a car can provide the flexibility needed to move to a better location if weather conditions change or if access to your chosen site is more difficult than expected. Ensure you have enough fuel, water, and provisions, especially if you're travelling to a remote area.

If you're planning to observe the eclipse in a foreign country, check the availability of public transportation, hire guides if necessary, and have a backup plan in case of any changes in accessibility. Having a local guide can often be beneficial, particularly in areas where language barriers may exist or in remote locations that require specific knowledge for navigation.

5. Preparing for the Environment

The environmental conditions in your chosen viewing location are important factors to consider. If you're observing in a desert region, be prepared for significant temperature fluctuations between day and night. Dress in layers, bring plenty of water, and use sun protection during the day.

In tropical regions, high humidity and the possibility of sudden rain showers may require different preparations. Waterproof bags for your photography gear, as well as insect repellent, are good items to have on hand. Being fully prepared for the environment will ensure that your eclipse viewing is enjoyable and stress-free.

6. Safety and Equipment

Bringing appropriate safety gear for viewing the eclipse is a must. Ensure you have enough solar viewing glasses for everyone in your group and that your camera equipment is fitted with proper solar filters to avoid damaging your gear. Keep a checklist to ensure you have all necessary equipment, from camera batteries and tripods to remote releases and solar filters.

It's also essential to consider personal safety. If travelling to a remote location, let others know of your plans, bring a first aid kit, and have emergency contact information readily available. The remoteness of some eclipse locations means that cell phone coverage may be limited, so planning for emergency communication is advisable.

Understanding Eclipse Tourism

1. The Rise of Eclipse Tourism

Eclipse tourism has become increasingly popular over the last few decades, with people willing to travel significant distances to witness these rare celestial events. Unlike total eclipses, annular eclipses do not offer the complete darkening of the sky, but the ring of fire is still enough to attract thousands of enthusiasts who want to experience this visual spectacle. As a result, "eclipse chasers" have developed into a community of passionate travellers who seek out these occurrences wherever they happen in the world.

Many travel companies now offer dedicated eclipse tours, which often include transportation, accommodation, and guided viewing sessions with experienced astronomers. These packages are popular among those who prefer a well-organised trip without worrying about the logistical complexities involved in planning an eclipse viewing.

2. Benefits of Eclipse Tourism

For many regions, especially remote ones, eclipses bring a significant boost to tourism, which in turn benefits the local economy. Towns and villages along the path of annularity often experience an influx of visitors, leading to increased revenue for local businesses, from hotels and restaurants to markets and tour operators. For some regions, an eclipse represents a unique opportunity to promote tourism and showcase local culture.

Communities often take advantage of the increased tourism by organising events, festivals, and educational programs related to the eclipse. These events help visitors learn more about the science behind the eclipse, local culture, and the region's natural attractions, adding value to their experience.

3. Responsible Eclipse Tourism

While eclipse tourism can benefit local economies, it also comes with challenges, particularly concerning environmental impacts. Visitors must practice responsible tourism by minimising their footprint, respecting local customs, and ensuring that they do not leave litter behind, especially in remote or sensitive natural environments.

When travelling to view an eclipse, it is crucial to follow guidelines provided by local authorities, especially concerning protected areas. Some of the best viewing locations may be in environmentally sensitive areas, and it is vital to ensure that the event does not negatively impact local wildlife or ecosystems.

4. Combining Eclipse Viewing with Travel

Eclipse tourism often offers an excellent opportunity to combine the experience of the eclipse with a broader travel adventure. Many eclipse chasers choose destinations that provide additional attractions, such as cultural landmarks, natural wonders, or unique local experiences. For example, visitors to Chile for an eclipse might also explore the Atacama Desert, known for its otherworldly landscapes and dark skies perfect for stargazing.

In regions like East Africa, combining an eclipse with a safari tour can provide a unique way to experience the eclipse against the backdrop of incredible wildlife. Eclipse chasers visiting India might combine their viewing with a trip to historical sites like the Taj Mahal or the temples of Rajasthan, enriching the overall travel experience beyond just the eclipse event.

5. Public Viewing Events

In many locations, especially in urban areas, local governments or astronomical organisations arrange public viewing events for eclipses. These events are designed to make the eclipse accessible to everyone, with free solar glasses, educational talks, and telescope stations equipped with solar filters for safe viewing. These public gatherings create a communal atmosphere that enhances the experience of witnessing an annular eclipse, turning it into a shared celebration of science and the wonders of the natural world.

6. The Cultural Aspect of Eclipse Tourism

Eclipse viewing can be deeply intertwined with cultural experiences, particularly in regions where eclipses are seen as significant in local folklore or beliefs. In India, for example, eclipses have spiritual significance, and many rituals are performed during these events. Participating in or observing these cultural practices can provide a deeper understanding of how celestial events have shaped human culture.

In regions of Africa, witnessing an eclipse can be a communal experience that blends science with traditional stories. Engaging with local guides who share myths and legends associated with eclipses can add depth to the experience, transforming it from a mere visual spectacle into a cultural journey.

Planning a trip to witness an annular solar eclipse is an exciting opportunity for adventure, discovery, and wonder. By choosing the best location, ensuring proper travel logistics, and understanding the concept of eclipse tourism, you can enhance your experience of this extraordinary celestial event. Whether you are an eclipse chaser or a first-time viewer, seeing the "ring of fire" firsthand is an experience that combines the thrill of travel with the awe-inspiring beauty of the universe. Responsible planning and respect for local communities and environments will not only ensure a successful trip but also contribute positively to the places you visit, making your journey even more meaningful.

CHAPTER 8

Annular Eclipses and Their Effect on Nature

Annular solar eclipses, with their iconic "ring of fire" effect, have profound influences not just on human observers but also on the natural world. The temporary changes in sunlight, temperature, and the eerie dimming of day can trigger unusual reactions in both animals and plants, as well as impact atmospheric conditions and human behaviour. Exploring these effects provides insight into how even brief changes in solar radiation can alter the rhythms of nature and the emotional responses of living beings.

Observing Changes in Animals and Plants During an Eclipse

Eclipses have always fascinated scientists, particularly for their impact on the behaviour of animals and plants. The changes in light, temperature, and even atmospheric pressure that accompany an annular eclipse can elicit noticeable reactions from animals and plants that are used to predictable natural cycles.

1. Animal Behaviour During an Eclipse

During an annular solar eclipse, animals often exhibit behaviours that are typically associated with the onset of dusk. The sudden dimming of daylight can confuse diurnal animals that rely on light cues to regulate their daily activities.

- **Birds**: Birds are among the most visibly affected animals during an eclipse. Many species, particularly songbirds, are known to fall silent as the light diminishes, much like they do at sunset. Some birds may even begin to roost, settling into trees or bushes as if it were evening. After the eclipse ends and the light returns, birds can be seen resuming their normal activities, seemingly confused by the sudden reversal of "night."

- **Insects**: Insects such as crickets, which normally start chirping at twilight, may begin their evening chorus during the peak of an eclipse, mistaking the dim light for dusk. Bees, which are highly sensitive to changes in light intensity, often return to their hives as the sunlight fades. The sudden reduction in light can cause a flurry of activity as bees head back to safety, only to emerge again once the eclipse is over.

- **Farm Animals**: Livestock such as cows and chickens also respond to the changing light. Cows may head toward their barns, thinking it's time for their evening routine, while chickens often return to their coops. These responses demonstrate the reliance of many animals on light levels to regulate their internal clocks and daily routines.

- **Nocturnal Animals**: Interestingly, nocturnal animals, such as owls or bats, may become active during the eclipse. Owls can be heard calling, and bats may start foraging, thinking that twilight has arrived early. This behaviour, though brief, is indicative of the confusion caused by the eclipse's artificial dusk.

2. Plant Responses to Eclipses

Plants are also sensitive to changes in light and temperature, though their responses are less immediately noticeable than those of animals. Plants regulate many processes, such as photosynthesis and the opening and closing of their flowers, based on the presence of sunlight.

- **Photosynthesis**: During an eclipse, the reduction in light leads to a temporary decline in photosynthesis. For most plants, this interruption is too brief to have a significant effect on their overall growth or health, but it does demonstrate how even short-term changes in sunlight can disrupt natural processes. In controlled studies, scientists have observed that chlorophyll activity decreases as light diminishes, only to increase again once the Sun reappears.

- **Flowering Plants**: Some plants that close their flowers at night, such as certain species of daisies or morning glories, may respond to the dimming light by partially closing their blooms. This reaction, however, is not as pronounced as the behavioural changes observed in animals, and most flowers resume their typical state once the eclipse has passed.

3. Aquatic Life

Aquatic animals, such as fish and frogs, have also been observed displaying changes in behaviour during an eclipse. Fish that are active during the day may reduce their activity, while nocturnal species might become more mobile. Frogs, known for their distinctive evening calls, may begin to croak as the light fades, mistaking the eclipse for nightfall.

Atmospheric Changes: Light, Temperature, and Shadows

An annular solar eclipse brings with it significant, albeit temporary, changes to the atmosphere. These changes can alter the physical environment in ways that are both visually stunning and scientifically fascinating.

1. Changes in Light

The most obvious atmospheric change during an annular eclipse is the reduction in sunlight. The light during an annular eclipse takes on an eerie quality—daylight seems to fade into something similar to early twilight, but with a distinctly different colour temperature. The ring of fire, while still bright, does not provide the same intensity of light as the full Sun, and this unique lighting has a noticeable effect on the environment.

The quality of the light changes as well, often becoming more diffuse and taking on a cooler, bluish hue. This is because the majority of the Sun's direct light is blocked by the Moon, leaving only the scattered rays that pass through the atmosphere. The reduction in sunlight also means that shadows become sharper and more defined. The contrast between light and shadow becomes more pronounced, which is one of the most visually striking effects of an eclipse.

2. Temperature Drops

During an annular solar eclipse, the reduction in solar radiation leads to a noticeable drop in temperature. Depending on the location and the length of the eclipse, the temperature can drop by several degrees Celsius. In some cases, the cooling effect can be enough to cause a shift in local wind patterns, as cooler air displaces warmer air.

The temperature drop is particularly pronounced in areas that are directly within the path of annularity. Observers often report a sudden chill as the eclipse progresses, which adds to the surreal atmosphere of the event. This temperature change, while temporary, provides a clear illustration of the Sun's influence on the Earth's environment and the delicate balance that sustains our climate.

3. Unusual Shadow Effects

One of the more fascinating effects of an annular eclipse is the appearance of unusual shadows. As the light from the Sun is reduced to a thin ring, small gaps in leaves or other objects create crescent-shaped patterns on the ground. These crescent shadows are a result of the pinhole effect, where each small gap acts as a miniature projector, casting an image of the partially obscured Sun onto the ground.

This effect can be seen by simply placing a piece of cardboard with small holes on the ground during the eclipse and watching as hundreds of tiny crescent shapes form on the surface. These unusual shadows are one of the most beautiful and unique aspects of an eclipse, offering a tangible representation of the celestial mechanics at play.

4. Changes in Wind Patterns

The temperature drop during an eclipse can also cause changes in local wind patterns. As the ground cools, the air above it also cools, creating areas of differing air pressure. This change can lead to gentle breezes or shifts in wind direction, as the cooler air moves to replace warmer air that is rising. These wind changes are usually subtle but can add to the eerie and unique atmosphere of an eclipse.

Eclipse Impact on Human Behaviour and Emotions

Eclipses have always fascinated humans, and throughout history, they have been regarded with a mix of awe, curiosity, and sometimes fear. The impact of an annular eclipse on human behaviour and emotions is significant, as such an event is a powerful reminder of the forces of nature and our place in the universe.

1. Emotional Impact

The visual spectacle of an annular eclipse can evoke a wide range of emotions in those who witness it. The gradual darkening of the sky, the drop in temperature, and the sudden appearance of the ring of fire can create a sense of awe and wonder. Many people report experiencing a profound sense of connection with nature during an eclipse, as the event serves as a reminder of the vastness of the universe and the delicate balance that sustains life on Earth.

For some, the experience can be spiritual or transformative, evoking feelings of humility and reverence. The unique combination of visual, emotional, and even physical sensations during an eclipse makes it a powerful event that often leaves a lasting impression on those who witness it.

2. Historical Fear and Superstition

Throughout history, eclipses were often viewed with fear and suspicion, particularly in ancient cultures that did not understand the celestial mechanics behind these events. Annular eclipses, with their mysterious ring of fire, were sometimes seen as omens of doom or as indications of divine displeasure. This fear was not limited to ancient times—many cultures in more recent history have also regarded eclipses as events to be feared or respected.

Even today, some people may feel anxious or uneasy during an eclipse, particularly if they are unfamiliar with the science behind it. The sudden darkening of the sky and the alteration of the natural order can trigger an instinctive response rooted in our evolutionary past—a reminder of how dependent we are on the stability of our environment.

3. Social and Collective Experience

In modern times, annular eclipses have become opportunities for community gatherings and collective experiences. The excitement of witnessing a rare natural event often brings people together, whether at organised public viewing events or more informal gatherings. The shared experience of observing an eclipse can foster a sense of community and shared wonder, as people from all walks of life come together to witness a phenomenon that transcends borders and cultures.

The social aspect of eclipse viewing has been greatly enhanced by modern technology. Social media platforms allow people to share their experiences, photos, and emotions in real time, turning an already significant event into a global phenomenon. The communal aspect of eclipse viewing is a stark contrast to the fear and isolation that eclipses once evoked in the past.

4. Impact on Human Activities

Eclipses can also have practical impacts on human activities, particularly those that are dependent on sunlight. For example, solar power generation is temporarily reduced during an eclipse, which can have implications for energy supply in regions that rely heavily on solar power. In some cases, energy providers need to adjust their output or use backup sources to compensate for the reduced solar input.

The temporary reduction in light and drop in temperature can also affect outdoor activities and events. Farmers, for instance, may notice changes in animal behaviour, while people working outdoors might need to take breaks during the eclipse due to the sudden cooling or dimming of light.

5. Psychological and Physical Reactions

Some people experience psychological or even physical reactions during an eclipse. The sudden change in light can cause a sense of disorientation, and the drop in temperature may lead to physical sensations like chills or goosebumps. For those who have never witnessed an eclipse before, the experience can be overwhelming, with a mix of excitement and slight unease.

The psychological effects of an eclipse are linked to its rarity and the dramatic alteration of familiar surroundings. The day turning into night and then back again can create a sense of being outside the normal flow of time, which can be both thrilling and unsettling. This emotional response is a testament to the deep connection humans have with their environment and the importance of routine natural cycles in our lives.

Annular eclipses, with their striking visual and environmental effects, provide a unique opportunity to observe the intricate interplay between the celestial and natural worlds. The responses of animals, plants, and humans to these events highlight how interconnected we all are with the cycles of light and dark, warmth and cooling, and how even a brief alteration in these patterns can provoke profound reactions. Whether experienced as an awe-inspiring spectacle, a communal event, or an occasion for introspection, an annular solar eclipse serves as a powerful reminder of the dynamic forces that govern our universe and the delicate balance that supports life on Earth.

CHAPTER 9
Annular Eclipses in Popular Culture

Annular solar eclipses, with their striking "ring of fire" visual effect, have inspired countless works of art, literature, and music over the years. The dramatic transformation of the Sun and the sense of wonder they evoke make eclipses a powerful symbol, often used to represent change, transformation, and moments of profound mystery. Below, we explore the influence of annular eclipses in popular culture, their appearances in films, books, and art, as well as their influence on music, literature, and symbolism.

Eclipses in Films, Books, and Art

Annular eclipses have long fascinated artists, writers, and filmmakers, serving as dramatic devices that evoke awe, tension, and even fear. They have become popular motifs in various forms of storytelling, symbolising pivotal moments or serving as visual metaphors for significant changes.

1. Eclipses in Films

In films, solar eclipses are often depicted as dramatic events that signify major shifts in a story or the unleashing of supernatural forces. The mysterious appearance of the ring of fire is visually arresting, providing a powerful backdrop for emotional or climactic scenes.

- **"Apocalypto" (2006)**: Mel Gibson's "Apocalypto" is a prime example of how an eclipse—albeit not specifically annular—has been used to influence the story's direction. In this film, the protagonist is saved from being sacrificed due to the timely occurrence of an eclipse, which is interpreted by the priests as an omen. Though the eclipse in the film may not specifically depict an annular event, the same principles of solar obscuration and divine intervention resonate with how annular eclipses have been portrayed in other cultural narratives.

- **"Ladyhawke" (1985)**: In the film "Ladyhawke," an eclipse plays a critical role in the resolution of the story. The phenomenon is used as a symbolic bridge between two different worlds, allowing the protagonists, who are cursed, to be together for a brief moment. Though the eclipse here is not specified as annular, its portrayal underscores how eclipses can represent change, moments of transition, and the meeting of seemingly impossible conditions.

- **"The Twilight Saga: Eclipse" (2010)**: While the specific type of eclipse is not depicted in the third film of the Twilight series, the title itself reflects the importance of celestial events as symbols of transformation and tension. The eclipse here symbolises the obscuring of one phase by another, reflecting the personal conflicts faced by the characters. Eclipses, including annular ones, are often used this way in storytelling—to symbolise the temporary covering or revealing of hidden truths.

2. Eclipses in Books and Literature

Annular solar eclipses also play a role in literature, where their awe-inspiring and mysterious nature often serves as a metaphor for transformation or revelation. Writers have used eclipses to explore themes such as the fragility of the human condition, the power of nature, and the unknown.

- **"A Connecticut Yankee in King Arthur's Court" by Mark Twain**: In this classic novel, an eclipse—specifically a total one, though the theme is still relevant to annular eclipses—allows the protagonist to impress and manipulate those around him by predicting the celestial event. Twain uses the eclipse to comment on knowledge, power, and superstition.

Similarly, annular eclipses, with their dramatic appearance, have been used by authors to underscore moments of suspense or pivotal shifts in character development.

- **"The Last Eclipse" by James Maxwell**: This novel, part of a fantasy series, incorporates an annular eclipse as a key event that marks the transition between different epochs within the story. The eclipse serves as a visual marker of change, foretelling shifts in power and the dawn of a new era. Maxwell's use of the eclipse highlights how such celestial events, with their dramatic and almost mystical nature, are often seen as harbingers of transformation.

- **"The Time Traveler's Wife" by Audrey Niffenegger**: In this story, while eclipses are not explicitly central to the plot, they are referenced in the broader theme of how time affects human relationships. The fleeting and rare quality of an eclipse is likened to precious moments that can be experienced only once, similar to the way in which the protagonists' time together is fleeting and precious.

3. Eclipses in Art

Artists have been fascinated by solar eclipses for centuries, often depicting them in their works to capture their visual beauty and the emotions they evoke. The ring of fire created by an annular eclipse is especially compelling for visual artists.

- **J.M.W. Turner**: Turner, known for his evocative landscapes and dramatic use of light, often painted skies dominated by powerful celestial events, including eclipses. While not all his works are directly related to annular eclipses, the sense of awe and dramatic interplay of light and shadow that characterises his paintings echoes the effect of an eclipse.

- **Contemporary Installations**: Modern artists have also used the eclipse as a theme in their work. Annular eclipses, with their distinct, visually stunning ring of light, are depicted in many forms—from abstract installations to multimedia pieces that explore themes of obscurity, transition, and enlightenment. The ring of fire effect, in particular, lends itself to exploration of the contrast between light and darkness, order and chaos.

The Influence of Eclipses on Music and Literature

Annular solar eclipses have had a notable influence on both music and literature, often inspiring works that capture the mystery, beauty, and sometimes the ominous quality of this celestial event.

1. Eclipses in Music

Musicians often draw on celestial events for inspiration, and eclipses have provided fertile ground for creating pieces that evoke feelings of awe, mystery, and transformation.

- **Pink Floyd's "Eclipse"**: Although Pink Floyd's song "Eclipse" from the album "The Dark Side of the Moon" is more about metaphors than a literal eclipse, it captures the same sense of completeness and cosmic connection that annular eclipses evoke. The song's themes of darkness, light, and unity align closely with the experience of witnessing an eclipse, where different elements of the cosmos align to create a rare and beautiful event.

- **Philip Glass's Opera "Eclipse"**: Minimalist composer Philip Glass created an opera inspired by celestial events, including solar eclipses. His work captures the gradual shifting and repetitive nature of an eclipse through musical motifs that build slowly, reflecting the way an eclipse progresses over time. The music's gradual crescendos mirror the way the light gradually fades and then returns, creating an auditory reflection of the visual spectacle.

- **Bon Iver's "Blood Bank"**: The lyrics of Bon Iver's "Blood Bank" reference the feeling of standing outside during an eclipse, capturing the shared human experience of witnessing something extraordinary in the sky. The song conveys a sense of wonder, as well as the intimacy of sharing such a rare moment with another person. This type of personal connection during an eclipse is often cited by those who witness annular events, where the sky's transformation creates a shared emotional experience.

2. Eclipses in Poetry and Literature

In literature, eclipses have been used by poets and writers as powerful symbols for a variety of themes, from the idea of fleeting beauty to the darker themes of obscurity and fear.

- **Sylvia Plath**: In her poem "The Moon and the Yew Tree," Plath alludes to an eclipse-like obscurity, blending images of the Moon and the dark branches of a tree to create a sense of foreboding. The imagery is reminiscent of an eclipse, where something beautiful becomes partially hidden, transforming into something mysterious or even threatening.

- **T.S. Eliot's "The Hollow Men"**: Eliot's exploration of darkness, ambiguity, and the feeling of being caught between two states resonates with the imagery of an eclipse. Though not explicitly referencing an annular eclipse, the sense of being suspended between light and dark, day and night, is a recurring theme in his work. Eclipses, including annular ones, provide a perfect metaphor for such liminality.

- **Emily Dickinson**: In her poems, Dickinson often explored celestial imagery. An eclipse can be seen as a symbol of the mysterious and the sublime—two recurring themes in her work. Dickinson's use of the natural world to explore human emotions aligns well with the impact that observing an eclipse can have on viewers, often leaving them with a renewed sense of awe and contemplation.

Annular Eclipses as Symbols of Change

Annular solar eclipses have long been symbols of change, transformation, and the cyclical nature of time. Their appearance in popular culture often emphasises their role as harbingers of important shifts or as markers of significant moments.

1. Symbolism in Mythology and Folklore

Historically, annular eclipses were often seen as omens, signalling major changes or significant events. This symbolism has carried over into modern popular culture, where eclipses often represent moments of transformation. In mythology, eclipses were frequently depicted as battles between light and dark, with the Sun being momentarily swallowed by a powerful force—only to re-emerge, transformed and renewed.

In many cultures, the eclipse was seen as a time when the boundaries between worlds became porous. This sense of transformation and fluidity is often used in popular culture to signify times of personal growth, change, or even chaos that precedes a new beginning. The annular eclipse, with its beautiful and unsettling ring of fire, embodies the idea that even in moments of darkness, there is light and beauty.

2. Eclipses as Catalysts for Change in Storytelling

In storytelling, annular eclipses are often used as pivotal moments that catalyse transformation. Whether they signify the beginning of a journey, the revealing of a hidden truth, or the climax of a conflict, eclipses have a way of focusing attention and highlighting the passage of time.

For example, an eclipse might mark the moment when a character gains new insight or when a prophecy is fulfilled. The dramatic visual transformation of an annular eclipse makes it an ideal metaphor for personal and collective transformation. The blocking of the Sun, followed by its reappearance, mirrors the way in which individuals or societies undergo periods of obscurity, only to emerge transformed and enlightened.

3. Annular Eclipses as Symbols of Unity

Another theme often associated with eclipses is unity. The alignment required for an eclipse to occur—a perfect lining up of the Sun, Moon, and Earth—symbolises harmony and balance. This alignment is sometimes used in popular culture to represent the idea that seemingly disparate elements can come together to create something beautiful and rare.

The ring of fire seen during an annular eclipse also represents continuity and wholeness, even in the face of apparent obstruction. This imagery is often used to convey messages of resilience and the cyclical nature of life. The temporary covering of the Sun, followed by the return of light, speaks to the idea that darkness is always followed by renewal, a theme that resonates deeply in both personal and collective narratives.

4. Personal and Spiritual Transformation

For many individuals, witnessing an annular eclipse can be a deeply personal experience, often described as a moment of introspection and transformation. The rare and fleeting nature of an eclipse reminds people of the temporary nature of life and the importance of appreciating beauty in its transience. This sense of change, of something extraordinary occurring that temporarily disrupts the ordinary flow of time, often finds its way into stories, art, and music as a metaphor for personal growth and transformation.

Annular eclipses have left a significant mark on popular culture, serving as powerful symbols of change, mystery, and beauty. Whether depicted in films, referenced in literature, or inspiring music and art, eclipses resonate with people as moments that transcend the ordinary, inviting reflection on the nature of the universe and our place within it. The ring of fire seen during an annular eclipse provides a vivid reminder of the interconnectedness of celestial events and human emotions, making these rare occurrences a potent source of inspiration and meaning.

CHAPTER 10
Future Annular Solar Eclipses

Annular solar eclipses, with their mesmerizing "ring of fire" effect, continue to captivate skywatchers worldwide. Predicting future eclipses, preparing for upcoming events, and examining the role of technology in enhancing our understanding of these celestial phenomena are essential aspects for anyone interested in eclipse observation. Below, we delve into the predictions for future annular solar eclipses, the preparations necessary for experiencing the next big event, and how technological advancements are poised to transform the way we observe these wonders.

Predicting Future Annular Eclipses: Dates and Locations

Annular eclipses occur when the Moon passes in front of the Sun but appears smaller due to its position further away in its elliptical orbit. This causes the bright ring of sunlight to remain visible, creating a unique spectacle. The path of annularity, which determines where on Earth an annular eclipse is visible in its entirety, changes with each event, as the orbits of the Earth and Moon continuously shift. Here is a list of some of the upcoming annular solar eclipses, including their dates and the locations where they will be visible.

1. Annular Solar Eclipse on October 2, 2024

- **Path of Annularity**: This annular eclipse will be visible across parts of South America, including Chile and Argentina, before moving across the Atlantic Ocean. This eclipse is expected to attract numerous observers, particularly because South America has become a popular destination for solar eclipses in recent years.

- **Optimal Viewing Locations**: Chile's Atacama Desert, known for its clear skies and dry climate, offers an excellent spot for observing this event. Argentina's Patagonia region is another good location, with its expansive open landscapes and typically clear October skies.

2. Annular Solar Eclipse on February 17, 2026

- **Path of Annularity**: This eclipse will traverse parts of Africa, including Angola, Zambia, and Mozambique, before moving across the Indian Ocean. The event will provide viewers in these regions with a chance to witness the spectacular ring of fire.

- **Optimal Viewing Locations**: Angola and Zambia have areas that offer unobstructed views of the horizon, ideal for observing the eclipse. Southern Africa generally experiences clear skies at this time of year, making it a promising destination for eclipse chasers.

3. Annular Solar Eclipse on January 26, 2028

- **Path of Annularity**: This annular eclipse will cross parts of Australia and New Zealand, providing an excellent opportunity for those in the Southern Hemisphere to witness the event.

- **Optimal Viewing Locations**: Northern Australia, including regions like the Northern Territory, will offer some of the best viewing spots. The deserts and open spaces of Australia provide a fantastic backdrop for eclipse photography, with the ring of fire appearing prominently against the clear Australian sky.

4. Annular Solar Eclipse on June 21, 2039

- **Path of Annularity**: This eclipse will pass through parts of Russia, Mongolia, and China. It will be one of the longest annular eclipses of the decade, providing ample time for observation.

- **Optimal Viewing Locations**: Mongolia's expansive plains and Russia's open countryside will provide excellent locations for viewing, where minimal light pollution allows for a clear view of the event.

Preparing for the Next Big Eclipse Event

Proper preparation is key to ensuring that an eclipse-viewing experience is memorable for all the right reasons. Whether it's your first time witnessing an annular eclipse or you're an experienced eclipse chaser, planning well in advance is crucial. Here's a guide to help you prepare for the next big annular eclipse event.

1. Choose the Best Location and Plan Early

Once you know the date and path of an upcoming annular eclipse, choose your viewing location as early as possible. The path of annularity is typically quite narrow, and spots within that path tend to get booked well in advance by other enthusiasts, scientists, and tourists. Use tools like NASA's eclipse maps or interactive websites like TimeandDate.com to find the best location along the path.

Consider factors such as accessibility, weather conditions, and safety. Desert areas, high-altitude locations, and places with historically clear weather are often ideal for eclipse viewing. Secure accommodations well in advance, as hotels in popular viewing locations can be fully booked months or even years before the event.

2. Ensure You Have Proper Viewing Equipment

Solar eclipses can be harmful to your eyes if viewed without adequate protection. Ensure that you have certified eclipse glasses that meet the ISO 12312-2 safety standard for direct solar viewing. If you plan to take photographs, you'll need solar filters for your camera and any other optical devices, such as telescopes or binoculars.

Practice using your equipment before the event. Make sure your solar filters fit securely, your camera settings are correct for solar photography, and all your gear is in good working order. Practicing in advance will help you troubleshoot any potential issues so that everything goes smoothly on the big day.

3. Monitor the Weather Forecast

Weather plays a significant role in the success of an eclipse-viewing trip. Cloud cover can completely obscure the view, so keeping a close eye on weather forecasts leading up to the eclipse day is crucial. Identify potential backup locations that are accessible within a reasonable distance, allowing you to change your viewing site if clouds are predicted.

4. Pack for Comfort and Safety

An eclipse can last several hours from the initial partial phases through to the end, so bring everything you need to be comfortable throughout the event. Pack water, snacks, sunscreen, hats, and extra layers of clothing. If you're in a remote area, ensure you have a first aid kit, plenty of water, and emergency supplies.

Remember to bring extra batteries and memory cards if you plan on photographing or filming the eclipse. You'll likely take far more shots than you initially expect, especially during the peak moments of the ring of fire.

5. Arrive Early and Scout Your Spot

Aim to arrive at your viewing site early—at least a day in advance if possible. This will give you time to scout the area, determine the best angle for viewing, and set up your equipment without feeling rushed. Being well-prepared will enhance your experience, allowing you to focus on the event itself rather than dealing with last-minute logistical issues.

6. Practice Safety Protocols

Make sure everyone in your group understands the importance of using proper eye protection during the eclipse. Never look directly at the Sun without proper solar viewing glasses, even during the annular phase, when the ring of fire is visible. Safety should be the number one priority for all eclipse viewers.

How Technological Advancements Might Change Eclipse Observations

Technological advancements have revolutionised how we experience and document eclipses. The next generation of eclipse observations is likely to be even more immersive, accessible, and data-rich, thanks to new tools, techniques, and platforms.

1. Drones and Aerial Imaging

One of the most exciting developments in recent years has been the use of drones for capturing aerial views of eclipses. Drones provide a unique perspective, allowing for sweeping views of the eclipse over a landscape. As drone technology continues to improve, with longer flight times and better stability, it's likely that future eclipse events will feature even more sophisticated aerial footage.

Drones can also be used to capture the effect of the eclipse on the environment—such as changing light conditions and animal behaviour—offering a broader context that isn't possible with ground-based cameras alone. This technology could make eclipse observation more dynamic and accessible, providing a new perspective that enhances understanding of the event.

2. Live Streaming and Virtual Reality

Live streaming has become a popular way to share the experience of a solar eclipse, especially for those who are unable to travel to the path of annularity. Advances in internet connectivity, coupled with high-quality cameras, have made it possible for people around the world to witness eclipses in real-time from the comfort of their homes.

The next step in this evolution is virtual reality (VR) and augmented reality (AR). VR technology could enable people to experience an eclipse as if they were physically present in the path of annularity. Imagine putting on a VR headset and being able to look up at the sky as the ring of fire forms above you, complete with the surrounding environmental effects, like changing light and temperature. Such experiences could make eclipses more accessible, particularly for those unable to travel to remote viewing locations.

3. Citizen Science and Data Collection

The rise of citizen science has changed the way we collect data during eclipses. Enthusiasts around the world, armed with smartphones and small telescopes, contribute valuable data about eclipses, including temperature changes, animal responses, and atmospheric effects. In the future, smartphone apps could provide platforms for coordinated data collection, allowing eclipse chasers to upload their observations in real-time to create comprehensive datasets for researchers.

These apps could also incorporate AI to help users set up their equipment, adjust settings for optimal eclipse photography, and even identify the best viewing locations. Such collaborative efforts would allow for the collection of massive amounts of data that could provide new insights into the impacts of eclipses on Earth.

4. Advances in Telescope Technology

Telescope technology continues to advance, making solar observation safer and more accessible for the general public. Affordable solar telescopes and filters are now available, allowing amateur astronomers to observe the Sun's surface features during an eclipse. Future advancements could include telescopes equipped with sophisticated solar tracking systems, allowing viewers to focus on enjoying the eclipse while the telescope automatically tracks the Sun's movement.

Another potential development is the use of adaptive optics—technology that compensates for atmospheric distortion in real time. Although currently used mainly in professional observatories, future portable versions of this technology could provide amateur eclipse chasers with sharper, more detailed images of the Sun during an eclipse, revealing features like solar flares and prominences in remarkable clarity.

5. Augmented Observation Experiences

Augmented reality (AR) applications may soon be used to enhance the eclipse-watching experience. For example, an AR-enabled smartphone or tablet could overlay real-time information onto the Sun during an eclipse, identifying different phases, showing the position of celestial bodies, or even providing educational commentary as the event progresses.

AR could also help educate the public about the mechanics of an eclipse, allowing viewers to understand the celestial alignments and orbital dynamics that produce these events. This technology has the potential to make eclipse viewing not only more interactive but also more educational, bridging the gap between spectacle and scientific understanding.

6. Advanced Weather Prediction and Cloud Analysis
One of the biggest challenges in eclipse viewing is the unpredictability of weather. Advances in meteorological technology and satellite-based weather prediction are expected to improve significantly in the coming years. Detailed cloud-cover prediction models will help eclipse chasers make last-minute decisions about where to position themselves, maximising their chances of a clear view.

Furthermore, new technologies, such as AI-based predictive models, could help analyse the likelihood of cloud formation and changes in weather patterns with high accuracy. This will enable more confident planning, allowing eclipse enthusiasts to choose the best viewing spots even when weather conditions are uncertain.

7. AI and Image Processing

Artificial intelligence is increasingly being used to enhance astrophotography. During an eclipse, AI-based tools can help process images in real-time, removing atmospheric interference and enhancing features such as solar prominences and the chromosphere. This allows photographers to produce clearer, more detailed images without spending hours in post-production.

In the future, AI might also be used to guide beginners in setting up their cameras, suggesting the best exposure settings based on current light conditions and assisting in tracking the Sun during the eclipse. These tools could democratise eclipse photography, allowing even those with little experience to capture high-quality images of the event.

Annular eclipses offer a spectacular window into the dynamic relationship between the Earth, Moon, and Sun. Predicting future eclipses allows enthusiasts to prepare for these extraordinary events, while technological advancements promise to make eclipse observation more immersive, informative, and accessible than ever before. From the use of drones and VR technology to the rise of citizen science and AI-enhanced astrophotography, the future of eclipse watching holds immense potential to deepen our connection to and understanding of these rare celestial events. As we continue to innovate, each new eclipse becomes not only a moment of natural wonder but also an opportunity to explore the boundaries of our technological capabilities.

CONCLUSION
Annular Solar Eclipses – An Extraordinary Celestial Phenomenon

Annular solar eclipses are among the most awe-inspiring celestial events visible from Earth, captivating people across cultures and continents for centuries. Characterised by the "ring of fire" effect, these eclipses offer a unique perspective on the interactions between our planet, its satellite, and the Sun. More than just visual spectacles, annular solar eclipses connect us with ancient traditions, inspire technological advancements, and encourage us to reflect on our place within the vast universe.

In understanding the complexity and significance of annular eclipses, one must consider not only their scientific underpinnings but also their impact on nature, culture, and human experience. These eclipses are moments when the predictable becomes profound—when everyday daylight is interrupted, and the Sun's radiance is briefly transformed into a golden halo. They remind us of the intricacies of celestial mechanics while also evoking an emotional and even spiritual response among viewers.

The Timeless Allure of Annular Eclipses

Throughout history, annular solar eclipses have played an important role in shaping cultural beliefs and scientific discoveries. Ancient civilisations, from the Maya to the Chinese, saw eclipses as moments of deep significance— omens that were often interpreted as messages from the heavens. They performed rituals and crafted myths that were passed down through generations, embedding eclipses within their cultural identities. The predictability of these events also led ancient astronomers to develop more accurate calendars and a deeper understanding of the movements of celestial bodies.

While modern science has demystified the mechanisms behind eclipses, allowing us to predict them with pinpoint accuracy, the sense of wonder remains unchanged. Today, eclipses are no longer feared as harbingers of doom; instead, they are celebrated as rare opportunities to witness the grandeur of the universe. Public viewing events, festivals, and educational programs accompany these celestial occurrences, allowing communities to come together and marvel at the alignment of the Earth, Moon, and Sun.

The annular eclipse, with its striking ring of fire effect, holds a particularly powerful visual and symbolic allure. Unlike total eclipses, where the Sun is completely obscured, annular eclipses retain a glow of sunlight, representing a bridge between light and darkness, between the ordinary and the extraordinary. This characteristic makes annular eclipses not only a feast for the eyes but also rich with metaphorical meaning—often seen as symbols of endurance, unity, and the cyclical nature of time.

Annular Eclipses and Their Impact on Nature

The natural world, too, is affected by these eclipses, demonstrating the interconnectedness of all living things with the rhythms of the cosmos. Animals that rely on light to determine their daily routines often react in surprising ways during an eclipse, mistaking the brief dimming of the Sun for the onset of dusk. Birds cease their singing, bees return to their hives, and nocturnal animals may briefly awaken. These reactions offer a glimpse into the instinctive responses that govern animal behaviour, showing how closely tied their lives are to the Sun.

Plants also react to the temporary reduction in sunlight, altering their photosynthesis rates or adjusting the opening and closing of flowers. These responses, though subtle, illustrate how even brief changes in solar radiation can affect biological processes. The reactions of both animals and plants during an annular eclipse serve as reminders of the delicate balance that sustains life on Earth and how dependent all living organisms are on consistent solar energy.

Beyond biological responses, annular eclipses also bring about noticeable changes in the atmosphere. The reduction in sunlight leads to a drop in temperature, creating a palpable chill that adds to the eerie atmosphere. The quality of light changes, shadows sharpen, and the landscape takes on an ethereal, almost dreamlike quality. These atmospheric changes are a visual reminder of the immense power of the Sun and the far-reaching effects of its light and heat on our environment.

Human Experience and Cultural Significance

The human experience of an annular solar eclipse goes beyond the scientific; it is deeply emotional, personal, and even spiritual for many. Witnessing an eclipse is often described as a profound experience—one that instills a sense of connection to the universe and a renewed awareness of the forces that govern our existence. The temporary transformation of the sky evokes feelings of wonder and humility, reminding us of our place within the vastness of the cosmos.

Eclipses also have a significant cultural impact. In popular culture, annular eclipses are often depicted in films, books, and art as symbols of change, transformation, or the unknown. They are used as metaphors for moments of revelation, for the interplay between light and shadow, and for the beauty that can emerge from moments of obscurity. In literature, music, and visual art, eclipses represent a powerful motif that resonates with the human experience of navigating between light and darkness, between clarity and mystery.

The ring of fire seen during an annular eclipse serves as a potent symbol of resilience and continuity. Even in moments when the Sun appears partially obscured, the light remains—a reminder that darkness is often temporary, and light persists. This symbolism has deep meaning in both collective narratives and personal journeys, where eclipses become metaphors for challenges that are faced and overcome, and for the hope that remains even in times of uncertainty.

The Rise of Eclipse Tourism

The allure of eclipses has led to the rise of eclipse tourism, with people traveling across continents to witness these rare events. The unique experience of seeing an eclipse has become a motivator for adventure, with enthusiasts—often called "eclipse chasers"—planning their travels years in advance. Annular eclipses, with their iconic ring of fire effect, attract people not only for their scientific interest but also for the shared experience of witnessing something truly spectacular.

Eclipse tourism brings people together, creating a sense of community among individuals from diverse backgrounds who share a common passion. Viewing an eclipse alongside others enhances the experience, transforming it from a solitary moment into a collective celebration of the wonders of the universe. The rise of public viewing events, often accompanied by educational talks and telescope stations, helps demystify the phenomenon and inspires new generations to explore astronomy.

These gatherings are also beneficial for local economies, particularly in regions that are along the path of annularity. The influx of tourists provides a boost to local businesses, from hotels and restaurants to transportation and guided tours. For many communities, an eclipse represents not only a rare natural event but also an opportunity to showcase their culture and hospitality to a global audience.

Technological Advancements and the Future of Eclipse Observation

Technological advancements have dramatically transformed how we observe and understand annular solar eclipses. Innovations in telescope technology, drone imaging, and live-streaming platforms have made it possible for more people than ever to witness these events, even if they cannot be physically present in the path of annularity.

High-resolution imaging from drones and telescopes equipped with adaptive optics allows for clearer, more detailed observations of the Sun during an eclipse. Citizen science initiatives, enabled by smartphone apps and online platforms, have turned eclipse viewing into a collaborative effort, where enthusiasts contribute data and images to further scientific understanding. This collective approach has enriched our knowledge of eclipses, particularly regarding their atmospheric effects and impacts on wildlife.

Virtual reality and augmented reality technologies are poised to revolutionise the eclipse-viewing experience, making it possible for people to experience an eclipse as if they were standing directly beneath it, even if they are halfway around the world. These advancements will make eclipses more accessible, providing immersive experiences that can inspire wonder and curiosity in audiences of all ages.

Eclipses as Moments of Reflection

Beyond the technological and cultural significance, annular solar eclipses also provide moments of reflection for those who witness them. The experience of seeing the Sun transformed into a ring of fire is a reminder of the beauty of our universe and the intricate dance of celestial bodies that make such events possible. It is a reminder that we are part of something far greater than ourselves, and that the forces at work in our solar system affect all life on Earth.

For many, witnessing an eclipse is a once-in-a-lifetime experience that inspires a deeper interest in astronomy and science. It is an opportunity to learn more about the mechanics of our solar system, to understand how eclipses occur, and to appreciate the role of the Sun in sustaining life. Eclipses have a way of sparking curiosity, inspiring us to explore and learn more about the world around us, and to seek out other celestial events that reveal the wonders of the universe.

The Legacy of Annular Eclipses

The legacy of annular solar eclipses is one of wonder, connection, and the pursuit of knowledge. From ancient civilisations that saw eclipses as divine omens to modern astronomers who use them to study the Sun's atmosphere, annular eclipses have always been moments of significance. They bring people together, inspire art and literature, challenge our understanding of nature, and remind us of the delicate balance that sustains life on our planet.

As we look to the future, we can anticipate even greater opportunities to experience these celestial events, aided by technological advancements that will make eclipse observation more immersive and informative. The wonder of the ring of fire will continue to draw people together, serving as a symbol of continuity in an ever-changing world. Annular eclipses, with their stunning beauty and profound symbolism, will continue to inspire awe, curiosity, and reflection for generations to come.

In conclusion, annular solar eclipses are not just astronomical phenomena—they are extraordinary moments that connect us to the universe, to each other, and to the natural world. They inspire creativity, curiosity, and community, reminding us of the intricate celestial dance that unfolds above our heads and the beauty that emerges when everything aligns. Whether experienced through the lens of a camera, in the pages of a book, or under the open sky, the ring of fire remains a testament to the wonders of our cosmos and our endless fascination with the mysteries it holds.

www.ingramcontent.com/pod-product-compliance
Lightning Source LLC
Chambersburg PA
CBHW050310230526
45471CB00005B/2114